MW01531577

LEAVES WE EAT

Katherine Rawson

Creating Young Nonfiction Readers

EZ Readers lets children delve into nonfiction at beginning reading levels. Young readers are introduced to new concepts, facts, ideas, and vocabulary.

Tips for Reading Nonfiction with Beginning Readers

Talk about Nonfiction

Begin by explaining that nonfiction books give us information that is true. The book will be organized around a specific topic or idea, and we may learn new facts through reading.

Look at the Parts

Most nonfiction books have helpful features. Our *EZ Readers* include a Contents page, an index, and color photographs. Share the purpose of these features with your reader.

Contents

Located at the front of a book, the Contents displays a list of the big ideas within the book and where to find them.

Index

An index is an alphabetical list of topics and the page numbers where they are found.

Photos/Charts

A lot of information can be found by "reading" the charts and photos found within nonfiction text. Help your reader learn more about the different ways information can be displayed.

With a little help and guidance about reading nonfiction, you can feel good about introducing a young reader to the world of *EZ Readers* nonfiction books.

Mitchell Lane
PUBLISHERS

2001 SW 31st Avenue
Hallandale, FL 33009
www.mitchelllane.com

First Edition, 2021.

Author: Katherine Rawson
Designer: Ed Morgan
Editor: Morgan Brody

Names/credits:
Title: Leaves We Eat / by Katherine Rawson
Description: Hallandale, FL :
Mitchell Lane Publishers, [2021]

Series: Plant Parts We Eat
Library bound ISBN: 978-1-58415-056-5
eBook ISBN: 978-1-58415-057-2

EZ Readers is an imprint of Mitchell Lane Publishers.

Photo credits: Freepik.com, Shutterstock

Contents

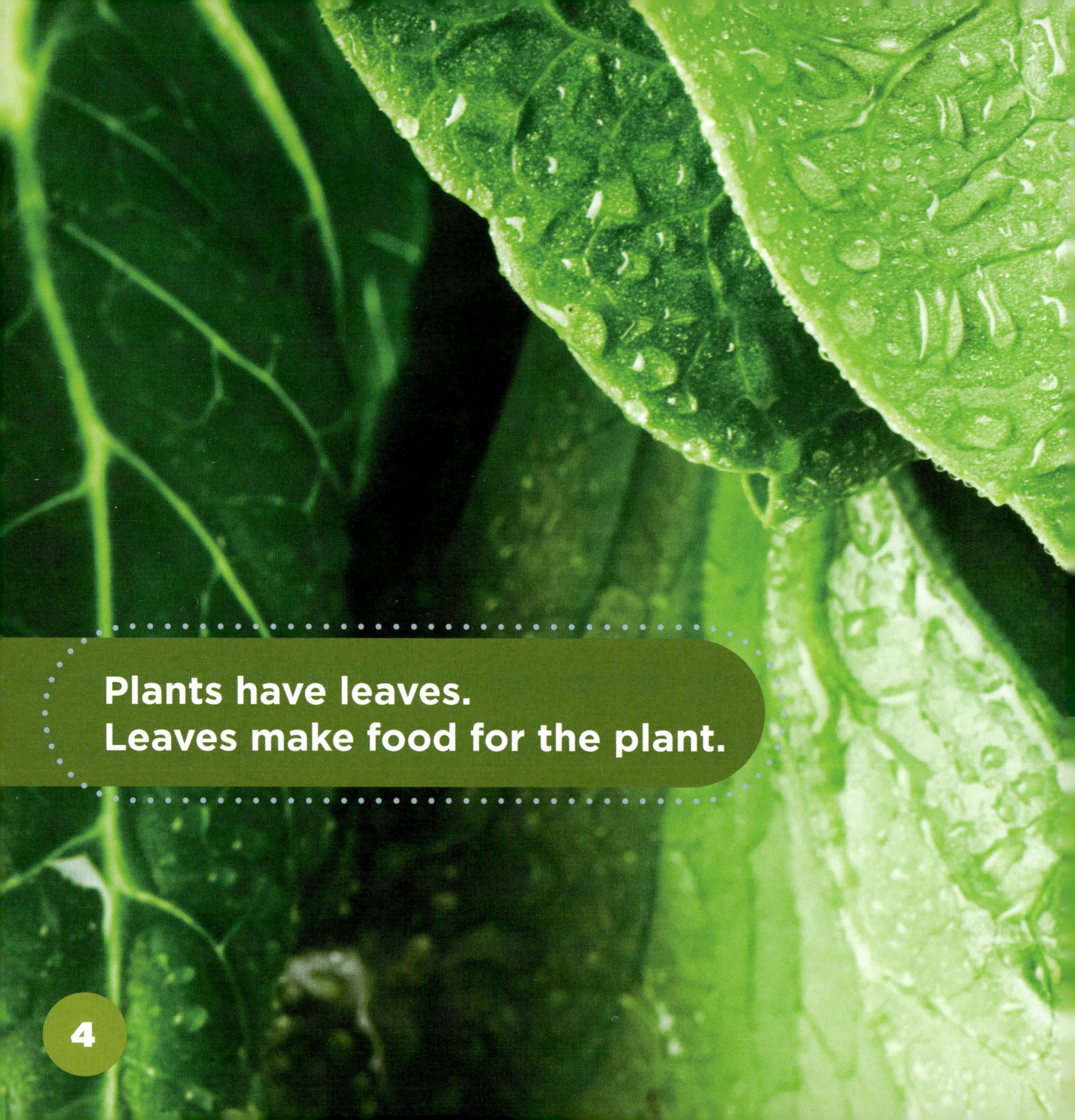

Plants have leaves.
Leaves make food for the plant.

They get water from the plant's roots.
They get **energy** from the sun.
They use the water and energy to
make food for the plant.

Leaves have different shapes and sizes.
Some leaves are good to eat.

Cabbage leaves are good to eat.
Cabbage leaves grow together
in a round **head**.
Cabbage can be green or purple.

DID YOU KNOW?
The heaviest cabbage in the world was grown in Alaska. It weighed almost 140 pounds!

It tastes good in salads.
It tastes good in soups and stews.

13

Spinach leaves are good to eat.
Spinach is dark green.
It has a lot of **vitamins**.

DID YOU KNOW?

Spinach has a lot of iron. Iron helps your blood stay healthy.

Spinach tastes good **raw**.
We eat it in salads.
Spinach tastes good cooked.
We eat it in soups and other **dishes**.

17

Lettuce leaves are good to eat.
Lettuce leaves grow together
in a head or a **bunch**.
We eat lettuce in salads and
sandwiches.

DID YOU KNOW?

There are may different kinds of lettuce. Dark green lettuce has more vitamins than light green lettuce.

PARSLEY

CHARD

Kale

Leaves are good to eat!

Beet Greens

Glossary

bunch
Leaves growing together in a group

dish
Prepared food

energy
Power

head
Leaves growing together in a round shape

raw
Not cooked

vitamins
Contents of food that we need for good health

Sources

https://www.bbc.com/bitesize/articles/z9gcdxs

http://www.houseplantsguru.com/the-function-of-leaves

https://www.livescience.com/51324-spinach-nutrition.html

https://www.nutritionletter.tufts.edu/issues/8_6/special-reports/How-Green-Is-Your-Salad_870-1.html

http://www.guinnessworldrecords.com/world-records/heaviest-cabbage

Further Reading

Web Pages:
https://extension.illinois.edu/gpe/case1/c1facts2c.html

Learn more about how leaves make food:
https://www.dkfindout.com/us/animals-and-nature/plants/how-plants-make-food/

Read more about plant parts we eat:
https://www.first-learn.com/food-from-plants.html

Books:
Photosynthesis
By Torrey Maloof
(Teacher Created Materials, May 20, 2015)

Seed to Plant
by Kristin Baird Rattini
(National Geographic Children's Books, 2014)

Index

About the Author

Katherine Rawson loves growing, cooking, and eating vegetables. She also loves writing. So, she thought it would be a great idea to write books about plants we eat. Her favorite leaf vegetable is spinach, and she likes to make spinach pizza.